PATCH GUIDE

U.S. Navy Ships and Submarines

TURNER PUBLISHING COMPANY

Author

The author was born in Enid, Oklahoma, enlisted in the U.S. Navy in 1957, and retired in 1979 as a Chief Aviation Machinists Mate. Since retirement he has remained in aerospace and is an executive for a major aircraft manufacturer.

Mr. Roberts is married to the former Judy White and resides in Long Beach, California. They have two children, a son Jay R. and a daughter Sherie Ann.

TURNER PUBLISHING COMPANY

Editor: Michael L. Roberts

Library of Congress
Catalog Card No.: 91-67691
ISBN: 978-1-63026-922-7

First Printing 1992

Preface

The ships crest is a customized emblem designed specifically for a particular unit. It is often displayed on plaques, patches, decals and other memorabilia. The crest builds pride in the crew and is a daily reminder that the crew belongs to the best ship in the U.S. Navy. In past years crests were designed by and for the crew members.

Although not approved by Navy regulations until the 1950s, ships have decorated their bridges and smokestacks with their crests since World War II.

Today, ship crests are authorized for "unofficial use only." Each element of the design has a specific meaning. Most crests are designed by professionals, with the design being approved by the Naval Board of Heraldry.

Table of Contents

TABLE OF CONTENTS (CONT'D)

ACKNOWLEDGEMENTS

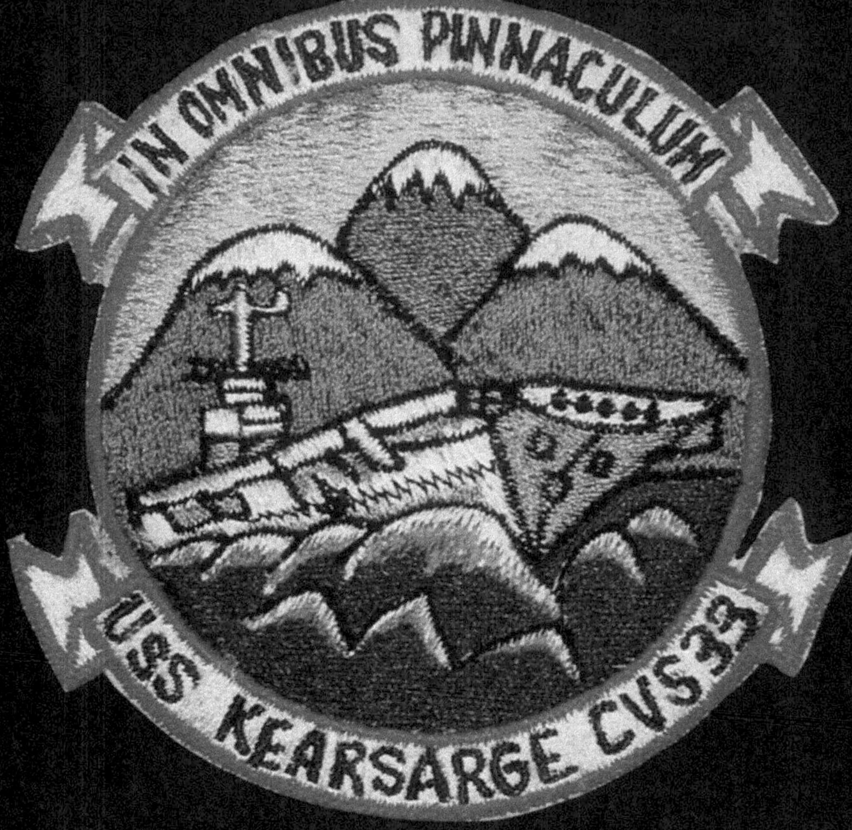

Credit is given to those individuals who provided valuable advice and untiring assistance without whose help this book would not have been published. My most sincere appreciation is given to:

Larry Adams, Photographer and friend.
Jeanette Weidner, Computer Consultant and friend.
Wayne Weidner, Computer Analyst and friend.
My son Jay R. for letting me steal time from him.
My wife Judy for her support and love.

The patches depicted in this book are from the author's personal collection.

UL: USS Dixie AD-14

UC: USS Prairie AD-15

UR: USS Cascade AD-16

ML: USS Piedmont AD-17

C: USS Sierra AD-18

MR: USS Yosemite AD-19

LL: USS Yosemite AD-19

LC:

LR: USS Yosemite AD-19
1974 Med. Cruise

UL: USS Yosemite AD-19
1974 Med. Cruise

UC: USS Hamul AD-20

UR: USS Arcadia AD-23

ML: USS Shenandoah AD-26

C: USS Isle Royal AD-29

MR: USS Tidewater AD-31

LL: USS Bryce Canyon AD-36

LC: USS Samuel Gompers AD-37

LR: USS Puget Sound AD-38

UL: USS Yellowstone AD-41

UC: USS Arcadia AD-42

UR: USS Cape Cod AD-43

ML: USS Shenandoah AD-44

C: USS Shasta AE-6

MR: USS Suribachi AE-21

LL: USS Firedrake AE-14

LC: USS Vesuvius AE-15

LR: USS Mount Katmai AE-16

UL: USS Great Sitkin AE-17
UC: USS Diamond Head AE-19
UR: USS Mauna Kea AE-22
ML: USS Nitro AE-23
C: USS Pyro AE-24
MR: USS Haleakala AE-25
LL: USS Haleakala AE-25 Cap Patch
LC: USS Kilauea AE-26
LR: USS Butte AE-27

UL: USS Santa Barbara AE-28

UC: USS Mount Hood AE-29

UR: USS Virgo AE-30

ML: USS Chara AE-31

C: USS Flint AE-32

MR: USS Shasta AE-33

LL: USS Mount Baker AE-34

LC: USS Kiska AE-35

LR: USS Arcturus AF-52

UL: USS Rigel AF-58
UC: USS Vega AF-59
UR: Los Alamos AFDB-7

ML: Competent AFDM-6
C: Sustain AFDM-7
MR: Resolute AFDM-10

LL: Steadfast AFDM-14
LC: USS Mars AFS-1
LR: USS Sylvania AFS-2

UL: USS Niagra Falls AFS-3

UC: USS White Plains AFS-4

UR: USS Concord AFS-5

ML: USS San Diego AFS-6

C: USS San Jose AFS-7

MR: USS Compass Island AG-153

LL: USS Assurance AG-521

LC: USS Glover AGDE-1

LR: USS Point Loma AGDS-2

UL: USS Plainview AGEH-1

UC: USS Banner AGER-1

UR: USS La Salle AGF-3

ML: USS La Salle AGF-3
Persian Gulf Yacht Club

C: USS Coronado AGF-11

MR: USS Coronado AGF-11

LL: USS Glover AGFF-1

LC: USS Arlington AGMR-2

LR: USS Graham County AGP-1176

UL: USS Watchman AGR-16

UC: USS Oxford AGTR-1

UR: USS Liberty AGTR-5

ML: USS Sanctuary AH-17

C: USS Oberon AKA-14

MR: USS Mathews AKA-96

LL: USS Rankin AKA-103

LC: USS Altair AKS-32

LR: USS Butternut AN-9

UL: USS Cimarron AO-22

UC: USS Chemung AO-30

UR: USS Guadalupe AO-32

ML: USS Maccaponi AO-41

C: USS Neches AO-47

MR: USS Ashtabula AO-51

LL: USS Cacapon AO-52

LC: USS Chikaskia AO-54

LR: USS Alagash AO-97

UL: USS Caloosahatchie AO-98

UC: USS Canisteo AO-99

UR: USS Mispillion AO-105

ML: USS Waccamaw AO-109

C: USS Neosho AO-143

MR: USS Neosho AO-143

LL: USS Mississinewa AO-144

LC: USS Hassayampa AO-145

LR: USS Kawishiwi AO-146

UL: USS Truckee AO-147

UC: USS Ponchatoula AO-148

UR: USS Ponchatoula AO-148 Power and Light

ML: USS Cimarron AO-177

C: USS Monongahela AO-178

MR: USS Merrimack AO-179

LL: USS Willamette AO-180

LC: USS Platte AO-186

LR: USS Sacramento AOE-1

UL: USS Camden AOE-2

UC: USS Seattle AOE-3

UR: USS Detroit AOE-4

ML: USS Wichita AOR-1

C: USS Milwaukee AOR-2

MR: USS Kansas City AOR-3

LL: USS Savannah AOR-4

LC: USS Wabash AOR-5

LR: USS Kalamazoo AOR-6

UL: USS Roanoke AOR-7

UC: USS Hunter Liggett APA-14

UR: USS American Legion APA-17

ML: USS George Clymer APA-27

C: USS Cavalier APA-37

MR: USS Hansford APA-106

LL: USS Magoffin APA-199

LC: USS Telfair APA-210

LR: USS Navarro APA-215

UL: USS Noble APA-218

UC: USS Pickaway APA-222

UR: USS Cook APD-130

ML: USS Weiss APD-135

C: USS Vulcan AR-5

MR: USS Ajax AR-6

LL: USS Hector AR-7

LC: USS Jason AR-8

LR: USS Jason AR-8

UL: USS Amphion AR-13

UC: USS Cadmus AR-14

UR: USS Markab AR-23

ML: USS Grand Canyon AR-28

C: USS Neptune ARC-2

MR: USS Aeolus ARC-3

LL: USS Waterford ARD-5

LC: USS White Sands ARD-20

LR: USS San Onofre ARD-30

UL: USS Oak Ridge ARDM-1

UC: USS Alamogordo ARDM-2

UR: USS Shippingport ARDM-4

ML: USS Arco ARDM-5

C: USS Sphinx ARL-24

MR: USS Escape ARS-6
Cap Patch

LL: USS Preserver ARS-8

LC: USS Safeguard ARS-25

LR: USS Bolster ARS-38

UL: USS Hoist ARS-40

UC: USS Opportune ARS-41

UR: USS Reclaimer ARS-42

ML: USS Safeguard ARS-50

C: USS Grasp ARS-51

MR: USS Salvor ARS-52

LL: USS Grapple ARS-53

LC: USS Fulton AS-11

LR: USS Sperry AS-12

UL: USS Bushnell AS-15

UC: USS Howard W. Gilmore AS-16

UR: USS Nereus AS-17

ML: USS Orion AS-18

C: USS Proteus AS-19

MR: USS Proteus AS-19

LL: USS Hunley AS-31

LC: USS Holland AS-32

LR: USS Holland AS-32

UL: USS Holland AS-32

UC: USS Simon Lake AS-33

UR: USS L.Y. Spear AS-36

ML: USS Canopus AS-34

C: USS Dixon AS-37

MR: USS Emory S. Land AS-39

LL: USS Frank Cable AS-40

LC: USS McKee AS-41

LR: USS McKee AS-41
W3 Division

UL: USS Florikan ASR-9

UC: USS Greenlet ASR-10

UR: USS Kittiwake ASR-13

ML:

C:

MR:

LL: USS Petrel ASR-14

LC: USS Sunbird ASR-15

LR: USS Skylark ASR-20

UL: USS Pigeon ASR-21	**UC: USS Ortolan ASR-22**	**UR: USS Tillamook ATA-192**
ML: USS Catawba ATA-210	**C: USS Kiowa ATF-72**	**MR: USS Mataco ATF-86**
LL: USS Abnaki ATF-96	**LC: USS Chowanoc ATF-100**	**LR: USS Hitchiti ATF-103**

UL: USS Moctobi ATF-105

UC: USS Molala ATF-106

UR: USS Takelma ATF-113

ML: USS Luiseno ATF-156

C: USS Nipmuc ATF-157

MR: USS Papago ATF-160

LL: USS Utina ATF-163

LC: USS Edenton ATS-1

LR: USS Beaufort ATS-2

UL: USS Brunswick ATS-3

UC: USS Tallahatchie County AVB-2

UR: USS Norton Sound AVM-1

ML: USS Orca AVP-49

C: USS Lexington AVT-16

MR: USS North Carolina BB-55

LL: USS South Dakota BB-57

LC: USS Iowa BB-61

LR: USS New Jersey BB-62

UL: USS New Jersey BB-62

UC: USS New Jersey BB-62

UR: USS New Jersey BB-62, Peace-keeping Force Beirut, Lebanon

ML: USS New Jersey BB-62 1983-84 Beirut, Lebanon

C: USS New Jersey BB-62 Battle Group Romeo

MR: USS Missouri BB-63

LL: USS Missouri BB-63

LC: USS Missouri BB-63 1990 Rim Pac.

LR: USS Wisconsin BB-64

UL: USS Saint Paul CA-73

UC: USS Toledo CA-133

UR: USS Los Angeles CA-135

ML: USS Newport News CA-148

C: USS Newport News CA-148

MR: USS Boston CAG-1

LL: USS Northampton CC-1

LC: USS Wright CC-2

LR: USS Little Rock CG-4

UL: USS Oklahoma City CG-5

UC: USS Albany CG-10

UR: USS Chicago CG-11

ML: USS Columbus CG-12

C: USS Leahy CG-16

MR: USS Harry E. Yarnell CG-17

LL: USS Worden CG-18

LC: USS Dale CG-19

LR: USS Richard K. Turner CG-20

UL: USS Gridley CG-21

UC: USS England CG-22

UR: USS Halsey CG-23

ML: USS Reeves CG-24

C: USS Belknap CG-26

MR: USS Joseph Daniels CG-27

LL: USS Wainwright CG-28

LC: USS Jouett CG-29

LR: USS Horne CG-30

UL: USS Sterett CG-31

UC: USS William H. Standley CG-32

UR: USS William H. Standley CG-32

ML: USS Fox CG-33

C: USS Biddle CG-34

MR: USS Biddle CG-34

LL: USS Truxton CG-35

LC: USS Ticonderoga CG-47

LR: USS Yorktown CG-48

UL: USS Vincennes CG-49

UC: USS Vincennes CG-49 1988 Gulf Games

UR: USS Valley Forge CG-50

ML: USS Thomas S. Gates CG-51

C: USS Bunker Hill CG-52

MR: USS Mobile Bay CG-53

LL: USS Antietam CG-54

LC: USS Leyte Gulf CG-55

LR: USS Leyte Gulf CG-55 1989 Med. Cruise

UL: USS Leyte Gulf CG-55
Operation Desert Storm

UC: USS San Jacinto CG-56

UR: USS San Jacinto CG-56
Operation Desert Storm

ML: USS Lake Champlain CG-57

C: USS Philippine Sea CG-58

MR: USS Princeton CG-59

LL: USS Princeton CG-59
Operation Desert Storm

LC: USS Normandy CG-60

LR: USS Monterey CG-61

UL: USS Chancellorsville CG-62

UC: USS Cowpens CG-63

UR: USS Gettysburg CG-64

ML: USS Chosin CG-65

C:

MR: USS Long Beach CGN-9

LL: USS Truxton CGN-35

LC:

LR: USS California CGN-36

UL: USS South Carolina CGN-37

UC: USS Virginia CGN-38

UR: USS Texas CGN-39

ML: USS Mississippi CGN-40

C: USS Arkansas CGN-41

MR: USS Arkansas CGN-41
1991 Persion Gulf

LL: USS Raleigh CL-7

LC: USS Columbia CL-56

LR:

UL: USS Biloxi CL-80`

UC:

UR: USS Manchester CL-83

ML: USS Roanoke CL-145

C: USS Galveston CLG-3

MR: USS Little Rock CLG-4

LL: USS Springfield CLG-7

LC: USS Topeka CLG-8

LR: USS Saratoga CV-3

UL: USS Yorktown CV-5

UC: USS Wasp CV-7

UR: USS Franklin CV-13

ML: USS Ticonderoga CVS-14

C:

MR: USS Oriskany CV-34

LL: USS Midway CV-41

LC: USS Midway CV-41, 1988 Official Carrier of the Olympics

LR: USS Midway CV-41 Operation Desert Shield

UL: USS Midway CV-41
1990-91 Persian Gulf

UC:

UR: USS Midway CV-41
Operation Desert Storm

ML: USS Coral Sea CV-43

C:

MR: USS Midway CV-41
Operation Desert Storm

LL: USS Forrestal CV-59

LC: USS Forrestal CV-59
1988 North Arabian Sea

LR: USS Philippine Sea CV-47

UL: USS Forrestal CV-59
1989 USA-USSR Summit

UC: USS Saratoga CV-60

UR: USS Saratoga CV-60
Operation Desert Storm

ML: USS Saratoga CV-60
Red Sea Yacht Club

C: USS Ranger CV-61

MR: USS Ranger CV-61
C.V.S.M.

LL: USS Ranger CV-61
1989 West. Pac.

LC:

LR: USS Ranger CV-61
Operation Desert Shield

UL: USS Ranger CV-61
Operation Desert Storm

UC:

UR: USS Independence CV-62

ML: USS Independence CV-62
Med. Cruise

C:

MR: USS Independence CV-62
1990 West. Pac.

LL: USS Independence CV-62
1990 Iraq Pac.

LC: USS Independence CV-62
Operation Desert Shield

LR: USS Independence CV-62
Foreign Legion

UL: USS Kitty Hawk CV-63

UC:

UR: USS Kitty Hawk CV-63
1981 West. Pac.

ML: USS Constellation CV-64

C:

MR: USS Constellation CV-64
1987 Indian Ocean

LL: USS Constellation CV-64
1990 Around the Horn

LC: USS Constellation CV-64
Connie Does S. America

LR: USS America CV-66

UL: USS America CV-66

UC:

UR: USS America CV-66
1979 Med. Cruise

ML: USS America CV-66
1982 St. Thomas

C: USS America CV-66

MR: USS America CV-66
Operation Desert Shield

LL: USS America CV-66
Operation Desert Storm

LC: USS John F. Kennedy CV-67

LR: USS John F. Kennedy CV-67
Drug Interdiction Cruise

UL: USS John F. Kennedy CV-67
1990 Drug Cruise

UC: USS John F. Kennedy CV-67
Operation Desert Storm

UR: USS John F. Kennedy CV-67
Operation Desert Storm

ML: USS John F. Kennedy CV-67
Operation Desert Storm

C: USS John F. Kennedy CV-67
Operation Desert Storm

MR: USS Essex CVA-9
1954-55 Far East

LL: USS Randolph CVA-15

LC:

LR: USS Hancock CVA-19

UL: USS Hancock CVA-19

UC:

UR: USS Boxer CVA-21

ML: USS Bon Homme Richard CVA-31

C: USS Oriskany CVA-34

**MR: USS Oriskany CVA-34
1974 West. Pac.**

LL:

LC: USS Shangri-La CVA-38

LR:

UL: USS Franklin D. Roosevelt CVA-42

UC: USS Coral Sea CVA-43

UR: USS Saratoga CVA-60
1958 Med. Cruise

ML: USS Ranger CVA-61

C: USS Constellation CVA-64

MR:

LL: USS America CVA-66
1972-73 West. Pac.

LC:

LR: USS Enterprise CVAN-65

UL: USS America CVA-66
1974 Med. Cruise

UC:

UR: USS John F. Kennedy CVA-67

ML: USS John F. Kennedy CVA-67
Med. Cruise

C: USS Nimitz CVAN-68

MR: USS Leyte CVE-32

LL: USS Gilbert Islands CVE-107

LC: USS Badoeing Straight CVE-116

LR: USS Palau CVE-122

UL: USS Cabot CVL-28

UC: USS San Jacinto CVL-30

UR: USS Enterprise CVN-65

ML: USS Enterprise CVN-65
1976-77 West. Pac.

C: USS Enterprise CVN-65
1983 Shellback

MR: USS Enterprise CVN-65
1989-90 World Cruise

LL: USS Enterprise CVN-65
1989-90 Just Cruising

LC: USS Nimitz CVN-68

LR: USS Nimitz CVN-68
Olympic Games

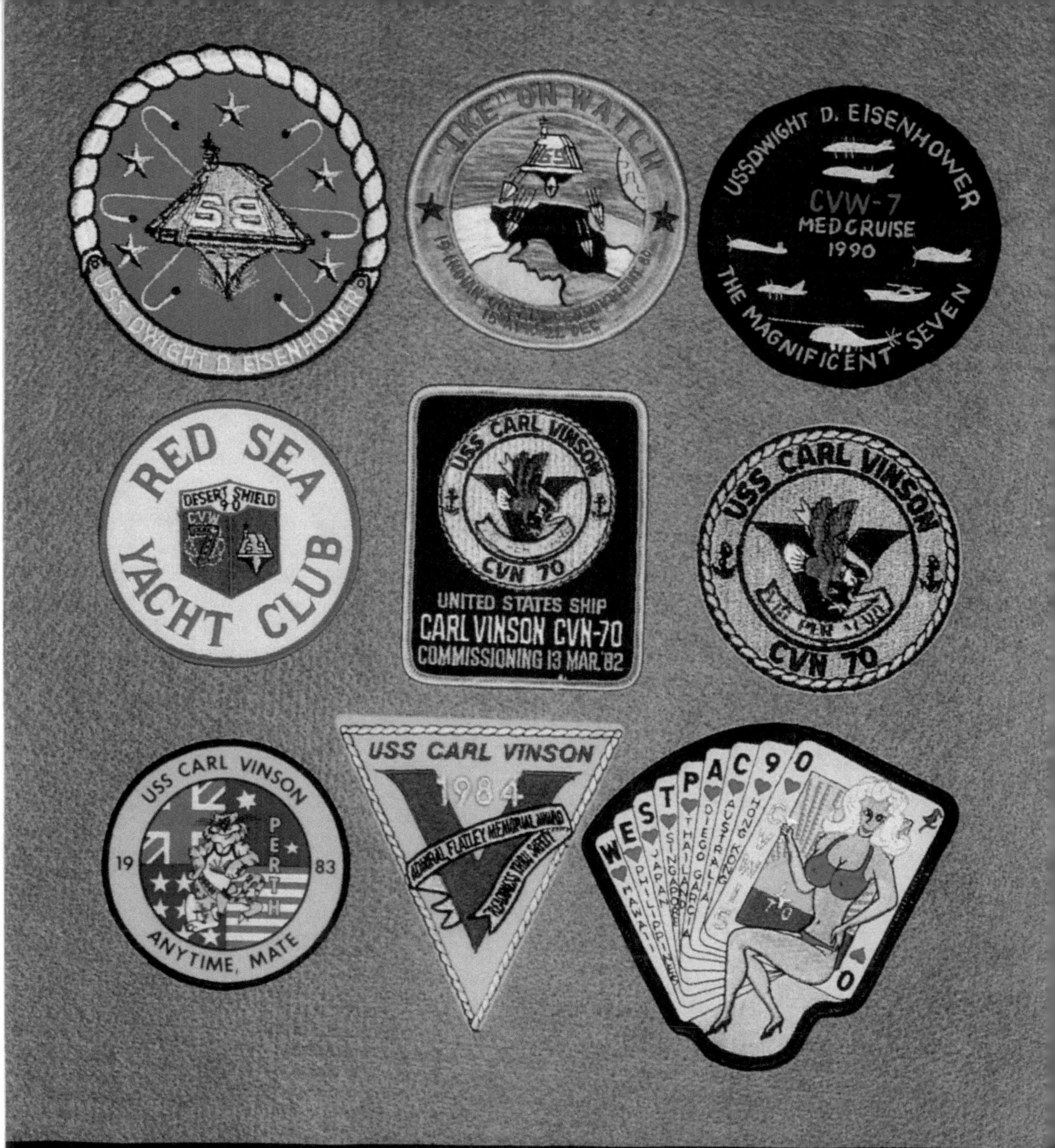

UL: USS Dwight David Eisenhower CVN-69

UC: USS Dwight David Eisenhower CVN-69, 1980 Indian Ocean

UR: USS Dwight David Eisenhower CVN-69, 1990 Med. Cruise

ML: USS Dwight David Eisenhower CVN-69, Red Sea Yacht Club

C: USS Carl Vinson CVN-70 1982 Commissioning

MR: USS Carl Vinson CVN-70

LL: USS Carl Vinson CVN-70 1983 Australia

LC: USS Carl Vinson CVN-70 1984 Admiral Flatly Award

LR: USS Carl Vinson CVN-70 1990 West. Pac.

UL: USS Carl Vinson CVN-70
Team Spirit

UC:

UR: USS Theodore Roosevelt CVN-71

ML: USS Theodore Roosevelt CVN-71
Operation Desert Storm

C:

MR: USS Theodore Roosevelt CVN-71
1991 Middle East Cruise

LL: USS Theodore Roosevelt CVN-71
Operation Desert Storm

LC:

LR: USS Theodore Roosevelt CVN-71
Operation Desert Storm

UL: USS Theodore Roosevelt CVN-71
Operation Desert Storm

UC: USS Abraham Lincoln CVN-72

UR: USS Abraham Lincoln CVN-72
Maiden Voyage Around the Horn

ML: USS Abraham Lincoln CVN-72
Desert Storm Victory Cruise

C: USS Essex CVS-9

MR: USS Yorktown CVS-10

LL: USS Intrepid CVS-11

LC: USS Hornet CVS-12

LR: USS Hornet CVS-12

UL: USS Randolph CVS-15

UC: USS Lexington CVS-16

UR: USS Wasp CVS-18

ML: USS Bennington CVS-20

C: USS Kearsarge CVS-33

MR: USS Antietam CVS-36

LL: USS Princton CVS-37

LC: USS Shangri-La CVS-38

LR: USS Lake Champlain CVS-39

UL: USS Lexington CVT-16

UC: USS Lexington CVT-16
1962-72 300,000 Landings

UR: CVN-65, CVN-70, CV-64
Nor. Pac. Glasnost

ML: USS Phelps DD-360

C:

MR: USS Rowan DD-405

LL: USS Chevalier DD-451

LC: USS Waller DD-466

LR: USS The Sullivans DD-537

UL: USS Twining DD-540
UC: USS Cowell DD-547
UR: USS Prichett DD-561

ML: USS Ross DD-563
C: USS Stoddard DD-566
MR: USS Wren DD-568

LL: USS Gansevoort DD-608
LC: USS Abbot DD-629
LR: USS Erben DD-631

UL: USS Hale DD-642

UC: USS Ingersoll DD-652

UR: USS Bearss DD-654

ML: USS Chauncey DD-667

C:

MR: USS Dortch DD-670

LL: USS Picking DD-685

LC: USS Remey DD-688

LR: USS Allen M. Sumner DD-692

UL: USS Ingraham DD-694

UC: USS Chas. S. Sperry DD-697

UR: USS Ault DD-698

ML: USS Waldron DD-699

C: USS Hank DD-702

MR: USS Borie DD-704

LL: USS Gainard DD-706

LC: USS Wm. R. Rush DD-714

LR: USS Wiltsie DD-716

UL: USS Theodore E. Chandler DD-717

UC: USS Hamner DD-718

UR: USS Barton DD-722

ML: USS De Haven DD-727

C: USS Lyman K. Swenson DD-729

MR: USS Collett DD-730

LL: USS Blue DD-744

LC: USS Southerland DD-745

LR: USS Samuel N. Moore DD-747

UL: USS Harry E. Hubbard DD-748

UC: USS Frank E. Evans DD-754

UR: USS Strong DD-758

ML: USS Lofberg DD-759

C: USS John W. Thomason DD-760

MR: USS Henley DD-762

LL: USS William C. Lawe DD-763

LC: USS Lloyd Thomas DD-764

LR: USS Keppler DD-765

UL: USS Zellars DD-777

UC: USS R. K. Huntington DD-781

UR: USS Gurke DD-783

ML: USS McKean DD-784

C: USS Henderson DD-785

MR: USS Richard B. Anderson DD-786

LL: USS James E. Kyes DD-787

LC: USS Hollister DD-788

LR: USS Eversole DD-789

UL: USS Shelton DD-790

UC: USS Benham DD-796

UR: USS Colhoun DD-801

ML: USS Gregory DD-802

C: USS Chevalier DD-805

MR: USS Higbee DD-806

LL: USS Corry DD-817

LC: USS New DD-818

LR: USS Holder DD-819

UL: USS Rich DD-820

UC: USS Johnston DD-821

UR: USS Johnston DD-821

ML: USS R.H. McCard DD-822

C: USS Samuel B. Roberts DD-823

MR: USS Carpenter DD-825

LL: USS Robert A. Owens DD-827

LC: USS Herbert J. Thomas DD-833

LR: USS Charles P. Cecil DD-835

UL: USS George K. Mackenzie DD-836

UC: USS George K. Mackenzie DD-836

UR: USS Power DD-839

ML: USS Glennon DD-840

C: USS Noa DD-841

MR: USS Fiske DD-842

LL: USS Warrington DD-843

LC: USS Bausell DD-845

LR: USS Ozbourn DD-846

UL: USS Robert Wilson DD-847

UC: USS R.E. Kraus DD-849

UR: USS Rupertus DD-851

ML: USS Charles H. Roan DD-853

C: USS Fred T. Berry DD-858

MR: USS McCaffery DD-860

LL: USS Vogelgesang DD-862

LC: USS Steinaker DD-863

LR: USS Harold J. Ellison DD-864

UL: USS Cone DD-866

UC: USS Stribling DD-867

UR: USS Brownson DD-868

ML: USS Arnold J. Isbell DD-869

C: USS Damato DD-871

MR: USS Hawkins DD-873

LL: USS Henry W. Tucker DD-875

LC: USS Rogers DD-876

LR: USS Dyess DD-880

UL: USS Furse DD-882

UC: USS N.K. Perry DD-883

UR: USS Floyd B. Parks DD-884

ML: USS Orleck DD-886

C: USS Meredith DD-890

MR: USS Forest Sherman DD-931

LL: USS Barry DD-933

LC: USS Davis DD-937

LR: USS Jonas Ingram DD-938

UL: USS Manley DD-940 | **UC: USS Dupont DD-941** | **UR: USS Bigelow DD-942**

ML: USS Blandy DD-943 | **C: USS Mullinix DD-944** | **MR: USS Hull DD-945**

LL: USS Edson DD-946 | **LC: USS Morton DD-948** | **LR: USS Richard S. Edwards DD-950**

UL: USS Turner Joy DD-951
UC: USS Spraunce DD-963
UR: USS Paul F. Foster DD-964
ML: USS Kinkaid DD-965
C: USS Hewitt DD-966
MR: USS Elliot DD-967
LL: USS Arthur W. Radford DD-968
LC: USS Peterson DD-969
LR: USS Caron DD-970

UL: USS David R. Ray DD-971

UC: USS Oldendorf DD-972

UR: USS John Young DD-973

ML: USS Comte De Grasse DD-974

C: USS Comte De Grasse DD-974
1983 Stanavforlant

MR: USS O'Brien DD-975

LL: USS Merrill DD-978

LC: USS Briscoe DD-977

LR: USS Stump DD-978

UL: USS Conolly DD-979 **UC: USS Moosbrugger DD-980** **UR: USS John Hancock DD-981**

ML: USS Nicholson DD-982 **C: USS John Rodgers DD-983** **MR: USS Leftwich DD-984**

LL: USS Cushing DD-985 **LC: USS Harry W. Hill DD-986** **LR: USS O'Bannon DD-987**

UL: USS Thorn DD-988

UC: USS Deyo DD-989

UR: USS Ingersoll DD-990

ML: USS Fife DD-991

C: USS Fletcher DD-992

MR: USS Hayler DD-997

LL: USS Conway DDE-507

LC: USS Cony DDE-508

LR: USS Fred T. Berry DDE-858

UL: USS Charles F. Adams DDG-2

UC: USS John King DDG-3

UR: USS Lawrence DDG-4

ML: USS Claude V. Ricketts DDG-5

C: USS Barney DDG-6

MR: USS Henry B. Wilson DDG-7

LL: USS Lynde McCormick DDG-8

LC: USS Towers DDG-9

LR: USS Sampson DDG-10

UL: USS Sampson DDG-10
Operation Desert Shield

UC: USS Sellers DDG-11

UR: USS Robison DDG-12

ML: USS Hoel DDG-13

C: USS Buchanan DDG-14

MR: USS Berkley DDG-15

LL: USS Joseph Strauss DDG-16

LC: USS Conyngham DDG-17

LR: USS Semmes DDG-18

UL: USS Tattnall DDG-19

UC: USS Goldsborough DDG 20

UR: USS Cockrane DDG-21

ML: USS Benjamin Stoddert DDG-22

C:

MR: USS Richard E. Byrd DDG-23

LL: USS Waddell DDG-24

LC:

LR: USS Waddell DDG-24

UL: USS Decatur DDG-31

UC:

UR: USS John Paul Jones DDG-32

ML:

C: USS Parsons DDG-33

MR:

LL: USS Somers DDG-34

LC: USS Mitscher DDG-35

LR: USS John S. McCain DDG-36

UL: USS Farragut DDG-37

UC: USS Luce DDG-38

UR: USS MacDonough DDG-39

ML: USS Coontz DDG-40

C:

MR: USS King DDG-41

LL: USS Mahan DDG-42

LC:

LR: USS Dahlgren DDG-43

UL: USS Dahlgren DDG-43
1991 Round The Horn

UC: USS William V. Pratt DDG-44

UR: USS Dewey DDG-45

ML: USS Preble DDG-46

C: USS Arleigh Burke DDG-51

MR: USS Kidd DDG-993

LL: USS Callaghan DDG-994

LC: USS Scott DDG-995

LR: USS Chandler DDG-996

UL: USS Eugene A. Greene DDR-711 | **UC: USS W.R. Rush DDR-714** | **UR: USS William M. Wood DDR-715**

ML: USS Myles C. Fox DDR-829 | **C: USS Goodrich DDR-831** | **MR: USS Charles P. Cecil DDR-835**

LL: USS Ernest G. Small DDR-838 | **LC: USS Vesole DDR-878** | **LR: USS Leary DDR-879**

UL: USS Huse DE-145
UC: USS Brough DE-148
UR: USS Wm. C. Miller DE-259

ML: USS Greenwood DE-679
C: USS Coates DE-685
MR: USS Marsh DE-699

LL: USS Kyne DE-744
LC: USS Weeden DE-797
LR: USS Lester DE-1022

UL: USS Bridget DE-1024 | UC: USS Bauer DE-1025 | UR: USS Hooper DE-1026

ML: USS Joseph K. Taussig DE-1030 | C: USS John R. Perry DE-1034 | MR: USS McMorris DE-1036

LL: USS Bronstein DE-1037 | LC: USS Edward McDonnell DE-1043 | LR: USS Davidson DE-1045

UL: USS Voge DE-1047
UC: USS Sample DE-1048
UR: USS Koelsch DE-1049

ML: USS Albert David DE-1050
C: USS O'Callahan DE-1051
MR: USS Knox DE-1052

LL: USS Roark DE-1053
LC: USS Gray DE-1054
LR: USS Connole DE-1056

UL: USS W.S. Sims DE-1059 **UC: USS Lang DE-1060** **UR: USS Patterson DE-1061**

ML: USS Whipple DE-1062 **C: USS Reasoner DE-1063** **MR: USS Lockwood DE-1064**

LL: USS Stein DE-1065 **LC: USS Marvin Shields DE-1066** **LR: USS Francis Hammond DE-1067**

UL: USS Vreeland DE-1068

UC: USS Bagley DE-1069

UR: USS Downes DE-1070

ML: USS Badger DE-1071

C: USS Blakely DE-1072

MR: USS Robert E. Peary DE-1073

LL: USS Harold E. Holt DE-1074

LC:

LR: USS Trippe DE-1075

UL: USS Fanning DE-1076

UC: USS Joseph Hewes DE-1078

UR: USS Bowen DE-1079

ML:

C:

MR:

LL: USS Paul DE-1080

LC:

LR: USS Miller DE-1091

UL: USS Elmer Montgomery DE-1082

UC: USS Cook DE-1083

UR: USS Donald B. Beary DE-1085

ML:

C:

MR:

LL: USS Brewton DE-1086

LC: USS Kirk DE-1087

LR: USS Jesse L. Brown DE-1089

UL: USS Thomas C. Hart DE-1092

UC: USS Truett DE-1095

UR: USS Valdez DE-1096

ML: USS Moinester DE-1097

C: USS Brooke DEG-1

MR: USS Ramsey DEG-2

LL: USS Schofield DEG-3

LC: USS Talbot DEG-4

LR: USS Richard L. Page DEG-5

UL: USS Falgout DER-324

UC: USS Brister DER-327

UR: USS Kretchmer DER-329

ML: USS Forster DER-334

C: USS Savage DER-386

MR: USS Haverfield DER-393

LL: USS Mitscher DL-2

LC: USS Farragut DLG-6

LR: USS Luce DLG-7

UL: USS MacDonough DLG-8

UC: USS King DLG-10

UR: USS Mahan DLG-11

ML: USS William V. Pratt DLG-13

C: USS Dewey DLG-14

MR: USS Preble DLG-15

LL: USS Harry E. Yarnell DLG-17

LC: USS England DLG-22

LR: USS Reeves DLG-24

UL: USS Belknap DLG-26

UC: USS Josephus Daniels DLG-27

UR: USS Wainwright DLG-28

ML: USS Jouett DLG-29

C: USS Horne DLG-30

MR: USS Wm. H. Standley DLG-32

LL: USS Fox DLG-33

LC:

LR: USS Biddle DLG-34

UL: USS Bainbridge DLGN-25
UC: USS Truxton DLGN-35
UR: Trieste II DSV-1

ML: Turtle DSV-3
C: Sea Cliff DSV-4
MR: Mystic DSRV-1

LL: Avalon DSRV-2
LC: USS Observation Island EAG-154
LR: USS Julius A. Furer FF-6

UL: USS Bronstein FF-1037

UC: USS McCloy FF-1038

UR: USS Garcia FF-1040

ML: USS Bradley FF-1041

C: USS McDonnell FF-1043

MR: USS Brumby FF-1044

LL: USS Davidson FF-1045

LC: USS Voge FF-1047

LR: USS Sample FF-1048

UL: USS Koelsch FF-1049

UC: USS Albert David FF-1050

UR: USS O'Callahan FF-1051

ML: USS Knox FF-1052

C: USS Roark FF-1053

MR: USS Gray FF-1054

LL: USS Hepburn FF-1055

LC: USS Connole FF-1056

LR: USS Rathburne FF-1057

UL: USS Meyerkord FF-1058

UC: USS W.S. Sims FF-1059

UR: USS Lang FF-1060

ML: USS Patterson FF-1061

C: USS Whipple FF-1062

MR: USS Reasoner FF-1063

LL: USS Lockwood FF-1064

LC: USS Stein FF-1065

LR: USS Marvin Shields FF-1066

UL: USS Marvin Shields FF-1066

UC: USS Francis Hammond FF-1067

UR: USS Vreeland FF-1068

ML: USS Bagley FF-1069

C: USS Downes FF-1070

MR: USS Badger FF-1071

LL: USS Blakely FF-1072

LC: USS Robert E. Peary FF-1073

LR: USS Harold E. Holt FF-1074

UL: USS Tripple FF-1075 | UC: USS Fanning FF-1076 | UR: USS Ouellet FF-1077

ML: USS Joseph Hewes FF-1078 | C: USS Bowen FF-1079 | MR: USS Paul FF-1000

LL: USS Aylwin FF-1081 | LC: | LR: USS Elmer Montgomery FF-1082

UL: USS Cook FF-1083

UC: USS McCandless FF-1084

UR: USS Donald B. Beary FF-1085

ML: USS Brewton FF-1086

C: USS Kirk FF-1087

MR:

LL: USS Jessie L. Brown FF-1089

LC: USS Ainsworth FF-1090

LR: USS Barbey FF-1088

UL: USS Miller FF-1091

UC: USS Thomas C. Hart FF-1092

UR: USS Capodanno FF-1093

ML: USS Pharris FF-1094

C: USS Truett FF-1095

MR: USS Valdez FF-1096

LL: USS Moinester FF-1097

LC:

LR: USS Glover FF-1098

UL: USS Brooke FFG-1

UC: USS Ramsey FFG-2

UR: USS Schofield FFG-3

ML: USS Talbot FFG-4

C: USS Richard L. Page FFG-5

MR: USS Julias A. Furer FFG-6

LL: USS Oliver Hazard Perry FFG-7

LC: USS McInerney FFG-8

LR: USS Wadsworth FFG-9

UL: USS Wadsworth FFG-9

UC: USS Wadsworth FFG-9

UR: USS Duncan FFG-10

ML:

C:

MR:

LL: USS Clark FFG-11

LC: USS George Philip FFG-12

LR: USS Samuel Eliot Morison FFG-13

UL: USS Sides FFG-14 | **UC: USS Estocin FFG-15** | **UR: USS Clifton Sprague FFG-16**

ML: USS John A. Moore FFG-19 | **C: USS Antrim FFG-20** | **MR: USS Flatley FFG-21**

LL: USS Fahrion FFG-22 | **LC: USS Lewis B. Puller FFG-23** | **LR: USS Jack Williams FFG-24**

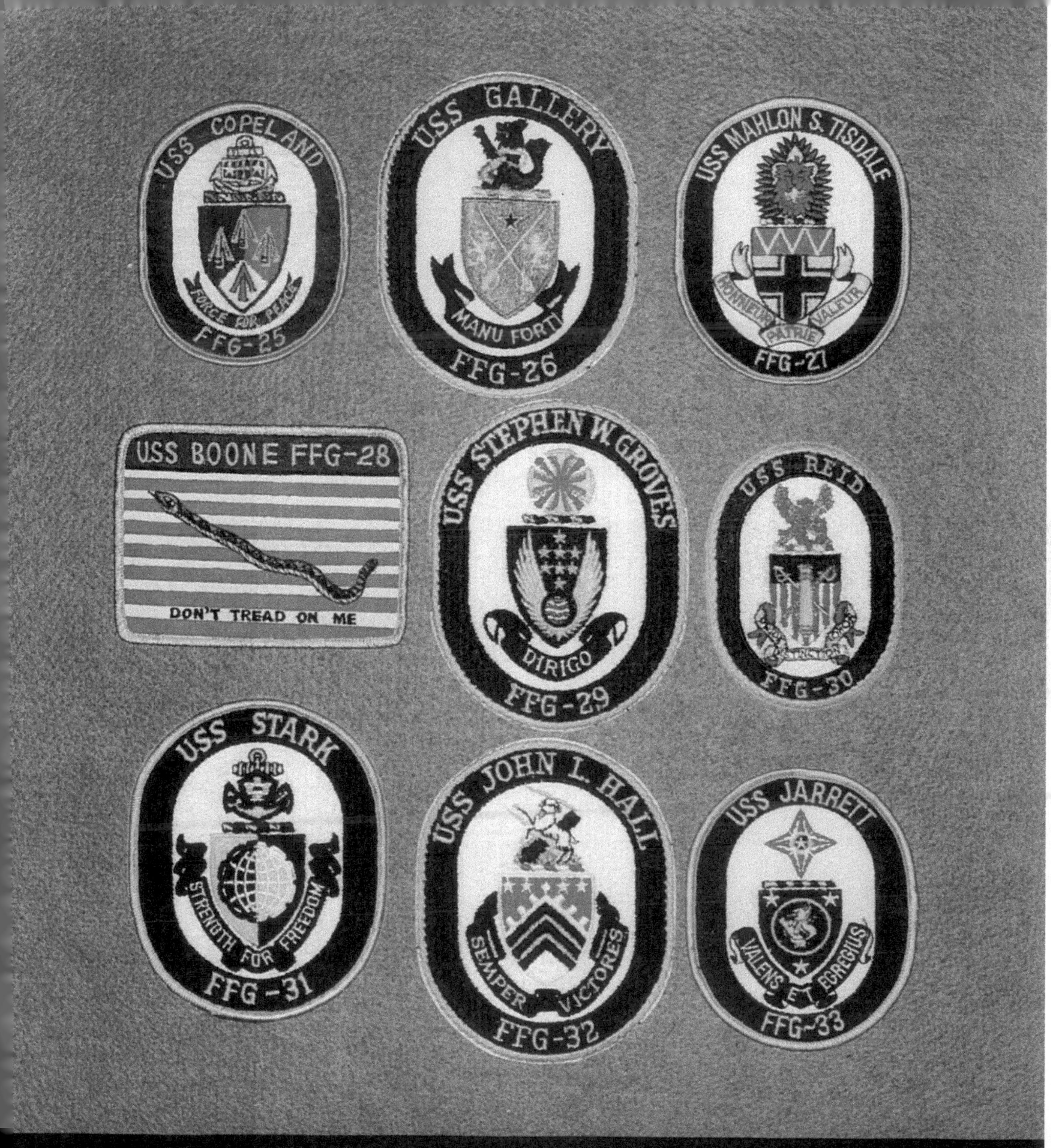

UL: USS Copeland FFG-25
UC: USS Gallery FFG-26
UR: USS Mahlon S. Tisdale FFG-27

ML: USS Boone FFG-28
C: USS Stephen W. Groves FFG-29
MR: USS Reid FFG-30

LL: USS Stark FFG-31
LC: USS John L. Hall FFG-32
LR: USS Jarrett FFG-33

UL: USS Aubrey Fitch FFG-34 **UC: USS Underwood FFG-36** **UR: USS Crommelin FFG-37**

ML: USS Curts FFG-38 **C: USS Doyle FFG-39** **MR: USS Halyburton FFG-40**

LL: USS McClusky FFG-41 **LC: USS Klakring FFG-42** **LR: USS Thach FFG-43**

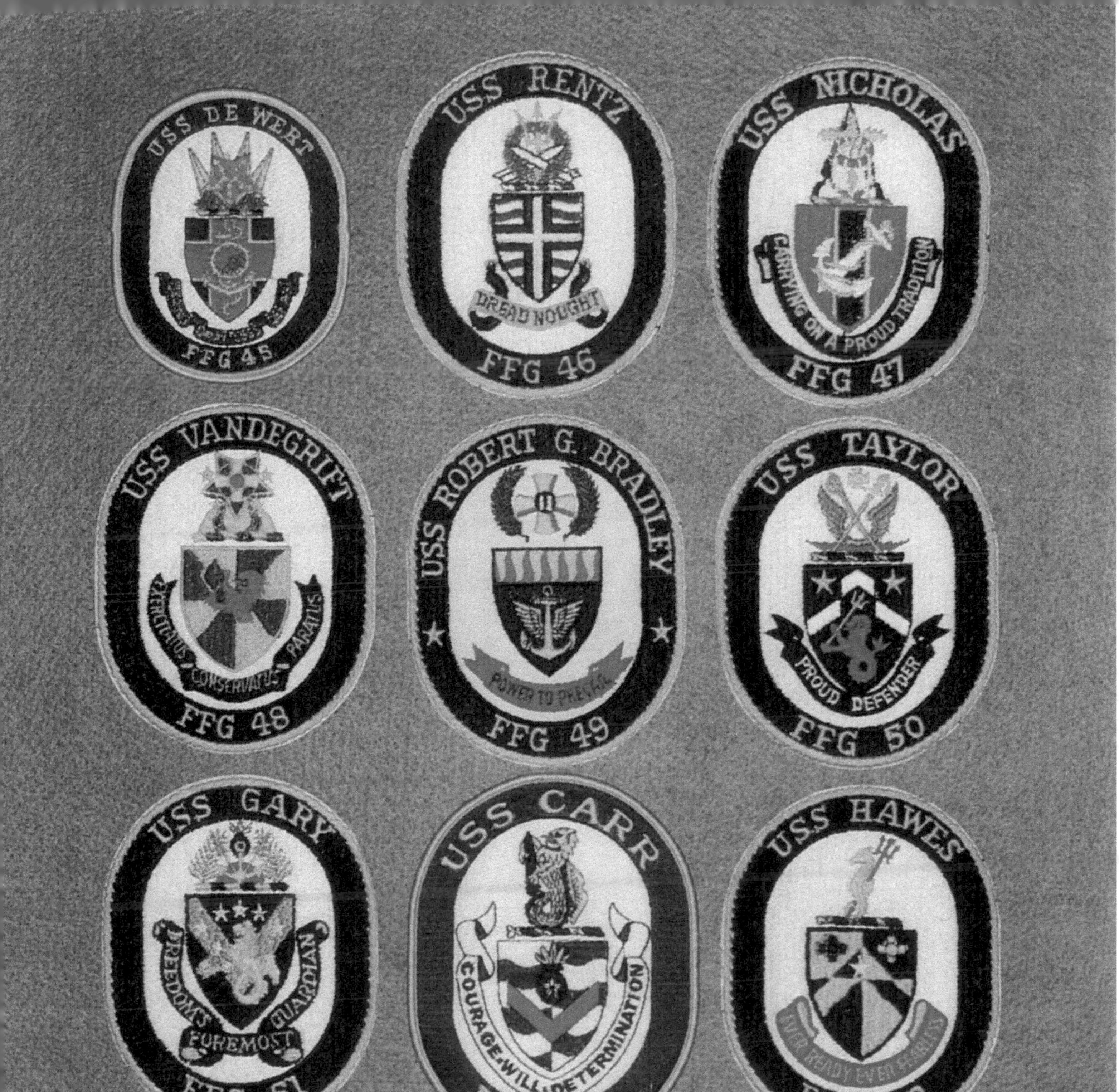

UL: USS De Wert FFG-45 **UC: USS Rentz FFG-46** **UR: USS Nicholas FFG-47**

ML: USS Vandegrift FFG-48 **C: USS Robert G. Bradley FFG-49** **MR: USS Taylor FFG-50**

LL: USS Gary FFG-51 **LC: USS Carr FFG-52** **LR: USS Hawes FFG-53**

UL: USS Ford FFG-54

UC:

UR: USS Ford FFG-54
Operation Desert Storm

ML. USS Elrod FFG-55

C: USS Reuben James FFG-57
Plankowner

MR: USS Simpson FFG-56

LL:

LC: USS Reuben James FFG-57
Engineering Dept., Plankowner

LR:

UL: USS Reuben James FFG-57

UC: USS Samuel B. Roberts FFG-58

UR: USS Samuel B. Roberts FFG-58
Operation Desert Shield

ML: USS Kauffman FFG-59

C: USS Rodney M. Davis FFG-60

MR: USS Ingraham FFG-61

LL: USS Constitution IX-21

LC:

LR: USS Elk River IX-501

UL: USS Blue Ridge LCC-19

UC: USS Mount Whitney LCC-20

UR: USS Tarawa LHA-1

ML: USS Saipan LHA-2

C: USS Saipan LHA-2
1985 Med. Cruise

MR: USS Belleau Wood LHA-3

LL: USS Nassau LHA-4

LC: USS Peleliu LHA-5

LR: USS Algol LKA-54

UL: USS Tulare LKA-112

UC: USS Tulare LKA-112 Cap Patch

UR: USS Charleston LKA-113

ML: USS Durham LKA-114

C: USS Mobile LKA-115

MR: USS St. Louis LKA-116

LL: USS El Paso LKA-117

LC: USS Paul Revere LPA-248

LR: USS Francis Marion LPA-249

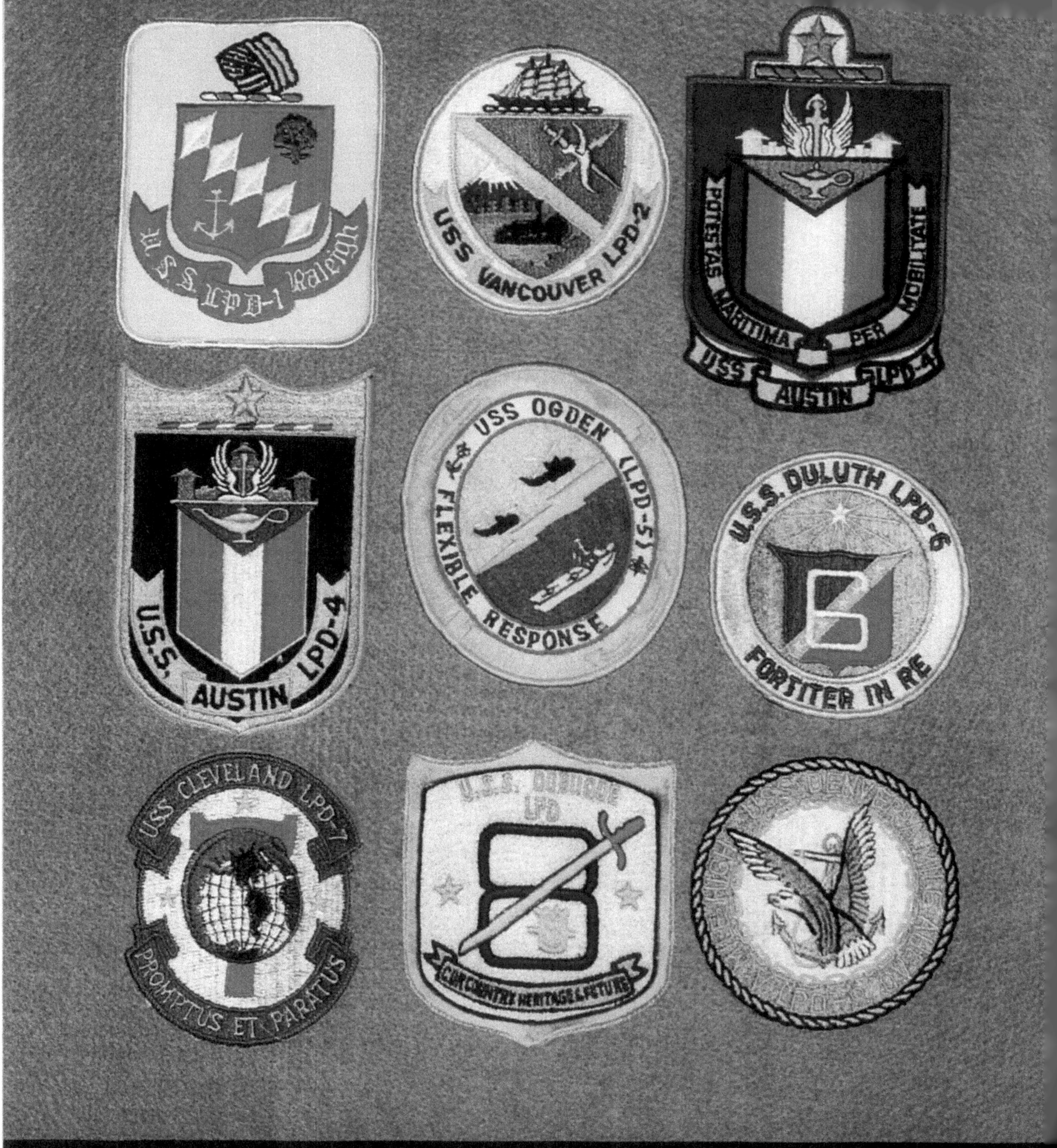

UL: USS Raleigh LPD-1

UC: USS Vancouver LPD-2

UR: USS Austin LPD-4

ML: USS Austin LPD-4

C: USS Ogden LPD-5

MR: USS Duluth LPD-6

LL: USS Cleveland LPD-7

LC: USS Dubuque LPD-8

LR: USS Denver LPD-9

UL: USS Juneau LPD-10

UC: USS Coronado LPD-11

UR: USS Shreveport LPD-12

ML: USS Nashville LPD-13

C: USS Trenton LPD-14

MR: USS Ponce LPD-15

LL: USS Iwo Jima LPH-2

LC: USS Okinawa LPH-3

LR: USS Okinawa LPH-3
Operation Desert Shield

UL: USS Guadalcanal LPH-7

UC: USS Valley Forge LPH-8

UR: USS Valley Forge LPH-8 Operation Desert Shield

ML: USS Guam LPH-9

C: USS Tripoli LPH-10

MR: USS Tripoli LPH-10 Operation Desert Shield

LL: USS New Orleans LPH-11

LC: USS Inchon LPH-12

LR: USS Inchon LPH-12 Operation Desert Storm

UL: USS Ashland LSD-1

UC: USS Epping Forest LSD-4

UR: USS Gunston Hall LSD-5

ML: USS Lind Enwald LSD-6

C: USS Oak Hill LSD-7

MR: USS Shadwell LSD-15

LL: USS Fort Mandan LSD-21

LC: USS Fort Marion LSD-22

LR: USS San Marcos LSD-25

UL: USS Thomaston LSD-28

UC: USS Plymouth Rock LSD-29

UR: USS Plymouth Rock LSD-29

ML: USS Fort Snelling LSD-30

C: USS Point Defiance LSD-31

MR: USS Spiegal Grove LSD-32

LL: USS Alamo LSD-33

LC: USS Hermitage LSD-34

LR: USS Monticello LSD-35

UL: USS Anchorage LSD-36

UC: USS Portland LSD-37

UR: USS Pensacola LSD-38

ML: USS Mount Vernon LSD-39

C: USS Fort Fisher LSD-40

MR: USS Whidbey Island LSD-41

LL: USS Germantown LSD-42

LC: USS Fort McHenry LSD-43

LR: USS Gunston Hall LSD-44

UL: USS Comstock LSD-45

UC: USS Duval County LST-758

UR: USS Garrett County LST-786

ML: USS Hunterdon County LST-838

C: USS Kemper County LST-854

MR: USS Litchfield County LST-901

LL: USS Luzerne County LST-902

LC: USS Page County LST-1076

LR: USS Snohomish County LST-1126

UL: USS Stark County LST-1134

UC: USS Stone County LST-1141

UR: USS Talahatchie County LST-1154

ML: USS Tioga County LST-1158

C: USS Vernon County LST-1161

MR: USS Waldo County LST-1163

LL: USS De Soto County LST-1171

LC:

LR: USS Suffolk County LST-1173

UL: USS Grant County LST-1174

UC: USS York County LST-1175

UR: USS Lorain County LST-1177

ML: USS Wood County LST-1178

C: USS Newport LST-1179

MR: USS Manitowoc LST-1180

LL: USS Sumter LST-1181

LC:

LR: USS Fresno LST-1182

UL: USS Peoria LST-1183

UC: USS Frederick LST-1184

UR: USS Schenectady LST-1185

ML: USS Cayuga LST-1186

C: USS Tuscaloosa LST-1187

MR: USS Saginaw LST-1188

LL: USS San Bernardino LST-1189

LC:

LR: USS Boulder LST-1190

UL: USS Racine LST-1191

UC: USS Spartanburg County LST-1192

UR: USS Fairfax County LST-1193

ML: USS La Moure County LST-1194

C: USS Barbour County LST-1195

MR: USS Harlan County LST-1196

LL: USS Barnstable County LST-1197

LC: USS Bristol County LST-1198

LR: USS Avenger MCM-1

UL: USS Sentry MCM-3

UC: USS Devastator MCM-6

UR: USS Catskill MCS-1

ML: USS Ozark MCS-2

C: USS Epping Forest MCS-7

MR: USS Frigate Bird MSC-191

LL: USS Vireo MSC-205

LC: USS Whippoorwill MSC-207

LR: USS Conflict MSO-426

UL: USS Constant MSO-427

UC: USS Constant MSO-427
Operation Desert Storm

UR: USS Dash MSO-428

ML: USS Detector MSO-429

C:

MR: USS Direct MSO-430

LL: USS Dominant MSO-431
Cap Patch

LC: USS Engage MSO-433

LR: USS Enhance MSO-437

UL: USS Excel MSO-439 | UC: USS Exultant MSO-441 | UR: USS Fearless MSO-442

ML: USS Fidelity MSO-443 | C: USS Fortify MSO-446 | MR: USS Illusive MSO-448

LL: USS Implicit MSO-455 | LC: USS Inflict MSO-456 | LR: USS Pluck MSO-464

UL: USS Salute MSO-470

UC: USS Conquest MSO-488

UR: USS Gallant MSO-489

ML: USS Pledge MSO-492

C: USS Affray MSO-511

MR: USS Cambria PA-36

LL: USS High Point PCH-1

LC: USS Hollidaysburg PCS-1385

LR: USS Gallup PG-85

UL: USS Antelope PG-86

UC: USS Crockett PG-88

UR: USS Marathon PG-89

ML: USS Tacoma PG-92

C: USS Grand Rapids PG-98

MR: USS Beacon PG-99

LL: USS Greenbay PG-101

LC: USS Flagstaff PGH-1

LR: USS Pegasus PHM-1

UL: USS Hercules PHM-2

UC: USS Taurus PHM-3

UR: USS Aquila PHM-4

ML: USS Aries PHM-5

C: USS Gemini PHM-6

MR: USS Nuclear Research Submarine NR-1

LL: USS Porpoise SS-172

LC: USS Permit SS-178

LR: USS Pollack SS-180

UL: USS Snapper SS-185

UC:

UR: USS Stingray SS-186

ML: USS Sturgeon SS-187

C: USS Spearfish SS-190
Battle Flag

MR: USS Greenling SS-213

LL: USS Silversides SS-236

LC:

LR: USS Angler SS-240

UL: USS Angler SS-240

UC:

UR: USS Bluegill SSK-242

ML: USS Bream SS-243

C: USS Cavalla SS-244

MR: USS Croaker SS-246

LL: USS Flasher SS-249

LC:

LR: USS Rasher SS-269

UL: USS Ray SSR-271

UC:

UR: USS Sawfish SS-276

ML: USS Tunny SS-282

C:

MR: USS Tunny APSS-282

LL: USS Bowfin SS-287

LC: USS Sabalo SS-302

LR: USS Ling SS-297

UL: USS Batfish SS-310

UC: USS Archerfish AGSS-311 Operation Seascan

UR: USS Perch APSS-313

ML: USS Barbero SSG-317

C:

MR: USS Barbero SSG-317

LL: USS Baya AGSS-318

LC:

LR: USS Becuna SS-319

UL: USS Becuna SS-319

UC: USS Bergall SS-320

UR: USS Blackfin SS-322

ML: USS Criman SS-323

C:

MR: USS Bullhead SS-332

LL: USS Carbonero SS-337

LC:

LR: USS Carbonero SS-337
1965 West. Pac.

UL: USS Carp SS-338

UC: USS Catfish SS-339

UR: USS Entemedor SS-340

ML: USS Clamagore SS-343

C: USS Cobbler SS-344

MR: USS Corporal SS-346

LL: USS Cubera SS-347

LC:

LR: USS Cusk SS-348

UL: USS Diodon SS-349

UC: USS Dogfish SS-350

UR: USS Halfbeak SS-352

ML: USS Hardhead SS-365

C: USS Icefish SS-367

MR: USS Lamprey SS-372

LL: USS Sandlance SS-381

LC:

LR: USS Parche SS-384

UL: USS Bang SS-385

UC: USS Bang SS-385

UR: USS Piranha SS-389

ML: USS Pomfret SS-391

C:

MR: USS Sterlet SS-392

LL: USS Redfish SS-395

LC:

LR: USS Ronquil SS-396

UL; USS Segundo SS-398

UC: USS Sea Cat SS-399

UR: USS Sea Devil SS-400

ML: USS Sea Dog SS-401

C: USS Sea Fox SS-402

MR: USS Atule SS-403

LL: USS Sea Owl SS-405

LC: USS Sea Robin SS-407

LR: USS Trepang SS-412

UL: USS Spot SS-413

UC: USS Stickleback SS-415

UR: USS Tiru SS-416

ML: USS Tigrone SS-419

C: USS Tirante SS-420

MR: USS Trutta SS-421

LL: USS Trumpetfish SS-425

LC: USS Tusk SS-426

LR: USS Corsair SS-435

UL: USS Argonaut SS-475

UC: USS Runner SS-476

UR: USS Conger SS-477

ML: USS Cutlass SS-478

C:

MR: USS Cutlass SS-478

LL: USS Diablo AGSS-479

LC:

LR: USS Requin SS-481

UL: USS Requin SSR-481

UC: USS Irex SS-482

UR: USS Odax SS-484

ML: USS Remora SS-487

C:

MR: USS Sarda SS-488

LL: USS Spinax SSR-489

LC: USS Gudgeon SS-507

LR: USS Amber Jack SS-522

UL: USS Dolphin SS-555
UC: USS Tang SS-563
UR: USS Trigger SS-564

ML: USS Wahoo SS-565
C: USS Trout SS-566
MR: USS Gudgeon SS-567

LL: USS Harder SS-568
LC: USS Sailfish SS-572
LR: USS Salmon SS-573

UL: USS Grayback LPSS-574

UC: USS Darter SS-576

UR: USS Growler SSG-577

ML: USS Barbel SS-580

C: USS Barbel SS-580

MR: USS Blueback SS-581

LL: USS Bonefish SS-582

LC: USS Bonefish SSR-582

LR: USS Mackeral SS-570

UL: USS George Washington SSBN-598

UC: USS George Washington SSBN-598 1959-85, Decommissioning Crew

UR: USS Patrick Henry SSBN-599

ML: USS Theodore Roosevelt SSBN-600

C: USS Robert E. Lee SSBN-601

MR: USS Abraham Lincoln SSBN-602

LL: USS Ethan Allen SSBN-608

LC: USS Sam Houston SSBN-609

LR: USS Thomas A. Edison SSBN-610

UL: USS John Marshall SSBN-611

UC: USS Lafayette SSBN-616

UR: USS Alexander Hamilton SSBN-617

ML: USS Thomas Jefferson SSBN-618

C: USS Thomas Jefferson SSBN-618 Purdums Pirates

MR: USS Thomas Jefferson SSBN-618 1963-1985, Decommissioning Crew

LL: USS Andrew Jackson SSBN-619

LC: USS John Adams SSBN-620

LR:

UL: USS James Monroe SSBN-622
UC: USS Nathan Hale SSBN-623
UR: USS Woodrow Wilson SSBN-624

ML: USS Henry Clay SSBN-625
C: USS Daniel Webster SSBN-626
MR: USS James Madison SSBN-627

LL: USS Daniel Boone SSBN-629
LC:
LR: USS John C. Calhoun SSBN-630

UL: USS U.S. Grant SSBN-631 UC: UR: USS Von Steuben SSBN-632

ML: USS Casimir Pulaski SSBN-633 C: USS Stonewall Jackson SSBN-634 MR: USS Sam Rayburn SSBN-635

LL: USS Nathanael Greene SSBN-636 LC: USS Benjamin Franklin SSBN-640 LR: USS Simon Bolivar SSBN-641

UL: USS Kamehameha SSBN-642

UC: USS George Bancroft SSBN-643

UR: USS Lewis & Clark SSBN-644

ML: USS James K. Polk SSBN-645

C: USS George C. Marshall SSBN-654

MR: USS Henry L. Stimson SSBN-655

LL: USS George Washington Carver SSBN-656

LC: USS Francis Scott Key SSBN-657

LR: USS Mariano G. Vallejo SSBN-658

UL: USS Will Rogers SSBN-659

UC: USS Ohio SSBN-726

UR: USS Michigan SSBN-727

ML: USS Florida SSN-728
Launch Crew

C: USS Florida SSBN-728

MR: USS Georgia SSBN-729

LL: USS Henry M. Jackson SSBN-730

LC: USS Alabama SSBN-731
Launch Crew

LR:

UL: USS Alabama SSBN-731

UC: USS Alaska SSBN-732
Launch Crew

UR: USS Alaska SSBN-732

ML: USS Alaska SSBN-732

C: USS Nevada SSBN-733
Launch Crew

MR: USS Nevada SSBN-733

LL: USS Tennessee SSBN-734
Launch Crew

LC:

LR: USS Tennessee SSBN-734

UL: USS Pennsylvania SSBN-735

UC:

UR: USS West Virginia SSBN-736

ML: USS Kentucky SSBN-737

C: USS Halibut SSGN-587

MR: USS Neptune SSN-525
Movie "Gray Lady Down"

LL: USS Nautilus SSN-571

LC: USS Seawolf SSN-575

LR: USS Skate SSN-578

UL: USS Swordfish SSN-579 | UC: USS Sargo SSN-583 | UR: USS Seadragon SSN-584

ML: USS Skipjack SSN-585 | C: USS Triton SSN-586 | MR: USS Triton SSRN-586

LL: USS Halibut SSN-587 | LC: USS Scamp SSN-588 | LR: USS Scorpion SSN-589

UL: USS Sculpin SSN-590

UC: USS Shark SSN-591

UR: USS Snook SSN-592

ML: USS Snook SSN-592
1981 Med. Run

C: USS Thresher SSN-593

MR: USS Permit SSN-594

LL: USS Permit SSN-594
1976 West . Pac. Patrol

LC: USS Plunger SSN-595

LR: USS Barb SSN-596

UL: USS Tullibee SSN-597

UC: USS Pollack SSN-603

UR: USS Haddo SSN-604

ML: USS Jack SSN-605

C: USS Tinosa SSN-606

MR: USS Dace SSN-607

LL: USS Guardfish SSN-612

LC: USS Flasher SSN-613

LR: USS Flasher SSN-613

UL: USS Greenling SSN-614

UC:

UR: USS Gato SSN-615

ML: USS Haddock SSN-621

C: USS Tecumseh SSBN-628

MR: USS Sturgeon SSN-637

LL: USS Whale SSN-638

LC: USS Whale SSN-638 Power and Light

LR: USS Tautog SSN-639

UL: USS Grayling SSN-646

UC: USS Pogy SSN-647

UR: USS Aspro SSN-648

ML: USS Sunfish SSN-649

C: USS Pargo SSN-650

MR: USS Queenfish SSN-651

LL: USS Puffer SSN-652

LC: USS Ray SSN-653

LR: USS Sand Lance SSN-660

UL: USS Lapon SSN-661

UC: USS Gurnard SSN-662

UR: USS Hammerhead SSN-663

ML: USS Sea Devil SSN-664

C: USS Guitarro SSN-665

MR: USS Hawkbill SSN-666

LL: USS Bergall SSN-667

LC: USS Spadefish SSN-668

LR: USS Seahorse SSN-669

UL: USS Finback SSN-670

UC: USS Narwhal SSN-671
Blue Crew

UR: USS Pintado SSN-672

ML: USS Flying Fish SSN-673

C: USS Flying Fish SSN-673
1989-90 Med. Run

MR: USS Trepang SSN-674

LL: USS Bluefish SSN-675

LC: USS Billfish SSN-676

LR: USS Drum SSN-677

UL: USS Archerfish SSN-678

UC: USS Silversides SSN-679

UR: USS William H. Bates SSN-680

ML: USS Batfish SSN-681

C:

MR: USS Tunny SSN-682

LL: USS Parche SSN-683

LC: USS Cavalla SSN-684

LR: USS Glenard P. Lipscomb SSN-685

UL: USS L. Mendel Rivers SSN-686

UC: USS Richard B. Russell SSN-687

UR: USS Richard B. Russell SSN-687
1990 Christmas Cruise, Blue Crew

ML: USS Richard B. Russell SSN-687
1991 Turban Tour

C:

MR: USS Los Angeles SSN-688

LL: USS Los Angeles SSN-688

LC: USS Baton Rouge SSN-689

LR: USS Philadelphia SSN-690

UL: USS Memphis SSN-691

UC: USS Omaha SSN-692

UR: USS Cincinnati SSN-693

ML: USS Groton SSN-694

C:

MR: USS Birmingham SSN-695

LL: USS Birmingham SSN-695
Battle "E" & Admin. "A"

LC:

LR: USS New York City SSN-696

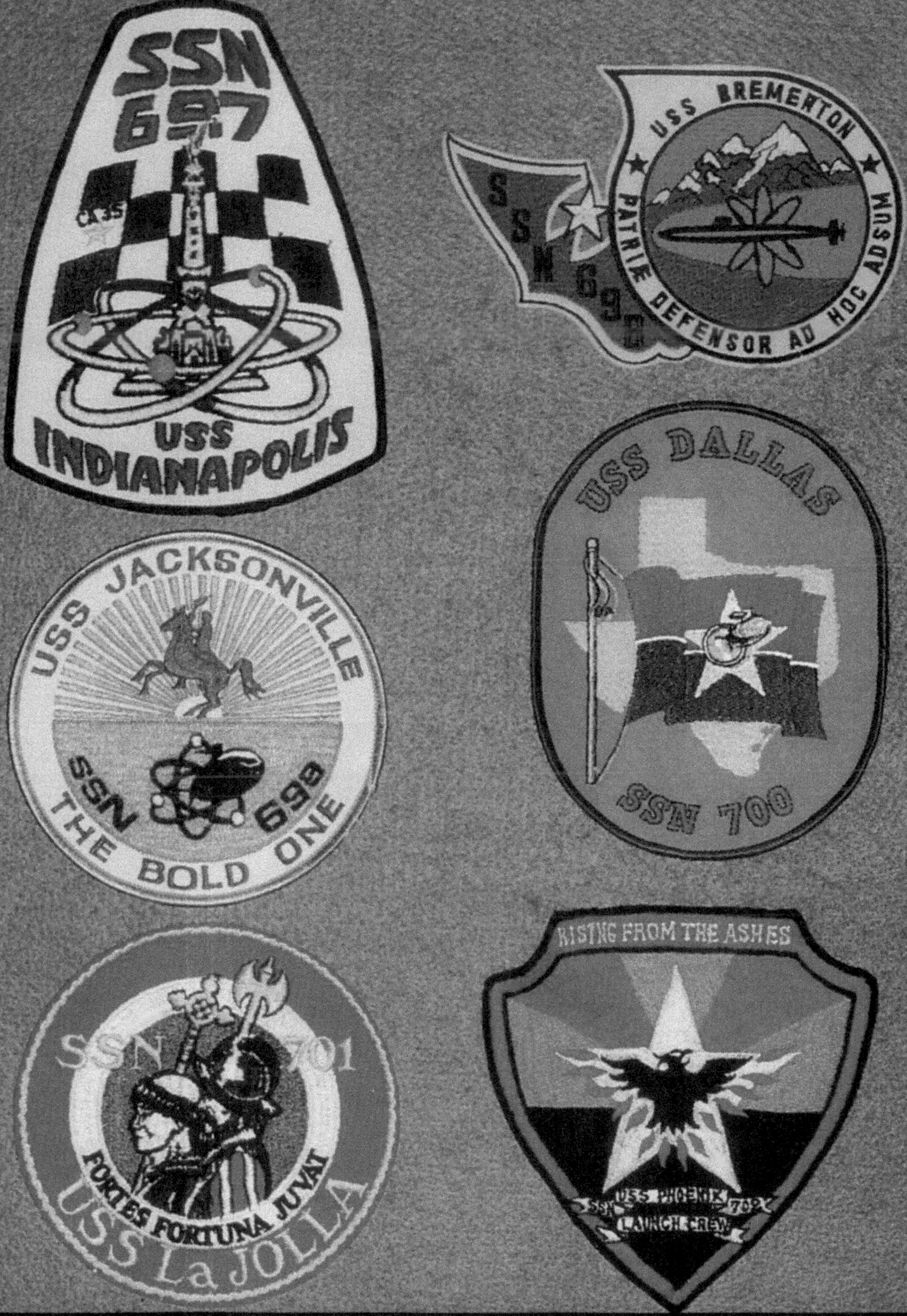

UL: USS Indianapolis SSN-697

UC:

UR: USS Bremerton SSN-698

ML: USS Jacksonville SSN-699

C:

MR: USS Dallas SSN-700

LL: USS La Jolla SSN-701

LC:

LR: USS Phoenix SSN-702

UL: USS Boston SSN-703
Launch Crew

UC:

UR: USS Boston SSN-703

ML: USS Baltimore SSN-704
Launch Crew

C:

MR: USS Baltimore SSN-704

LL: USS Corpus Christi SSN-705
Launch Crew

LC: USS Corpus Christi SSN-705

LR: USS City of Corpus Christi
SSN-705

UL: USS Albuquerque SSN-706

UC: USS Portsmouth SSN-707

UR: USS Minneapolis, St. Paul SSN-708

ML: USS Hyman G. Rickover SSN-709

C:

MR: USS Agusta SSN-710

LL: USS San Francisco SSN-711 Launch Crew

LC:

LR: USS San Francisco SSN-711

UL: USS Atlanta SSN-712
Launch Crew

UC: USS Atlanta SSN-712

UR: USS Atlanta SSN-712
1982 Autec

ML: USS Atlanta SSN-712
1983 Deployment

C: USS Houston SSN-713
Launch Crew

MR: USS Houston SSN-713

LL: USS Norfolk SSN-714
Plankowner

LC: USS Norfolk SSN-714
Launch Crew

LR: USS Norfolk SSN-714

UL: USS Buffalo SSN-715
Launch Crew

UC: USS Buffalo SSN-715

UR: USS Salt Lake City SSN-716
Launch Crew

ML: USS Salt Lake City SSN-716

C: USS Olympia SSN-717

MR: USS Honolulu SSN-718

LL: USS Providence SSN-719

LC: USS Pittsburgh SSN-720

LR: USS Chicago SSN-721
Cap Patch

UL: USS Chicago SSN-721

UC: USS Key West SSN-722

UR: USS Oklahoma City SSN-723

ML: USS Louisville SSN-724
Launch Crew

C: USS Louisville SSN-724

MR: USS Helena SSN-725
Launch Crew

LL: USS Helena SSN-725

LC: USS Newport News SSN-750

LR: USS San Juan SSN-751

UL: USS Pasadena SSN-752
1989 Commissioning

UC:

UR: USS Pasadena SSN-752

ML: USS Albany SSN-753

C: USS Topeka SSN-754
Pre-Commissioning Unit

MR: USS Topeka SSN-754

LL: USS Miami SSN-755

LC:

LR: USS Alexandria SSN-757

UL: USS Ashville SSN-758

UC:

UR: USS Jefferson City SSN-759

ML: USS Annapolis SSN-760 Launch Crew

C: USS Annapolis SSN-760

MR: USNS Rigel T-AF-58

LL: USNS Sirius T-AFS-8

LC: USNS Kingsport T-AG-164

LR: USNS S.P. Lee T-AG-192

UL: USNS Hayes T-AGOR-16

UC: USNS Vandenberg T-AGM-10

UR: USNS Vanguard T-AGM-19

ML: USNS Redstone T-AGM-20

C: USNS Bowditch T-AGS-21

MR: USNS Elisha Kent Kane T-AGS-27

LL: USNS Harkness T-AGS-32

LC: USNS Mercy T-AH-19

LR: USNS Mirfak T-AK-271

UL: USNS Norwalk T-AK-279

UC: USNS Marshfield T-AK-282

UR: USNS Mercury T-AKR-10

ML: USNS Jupiter T-AKR-11

C: USNS Marias T-AO-57

MR: USNS Waccamaw T-AO-109

LL: USNS Henry J. Kaiser T-A0-187

LC: USNS Neptune T-ARC-2

LR: USNS Aeolus T-ARC-3

UL: USS George Eastman YAG-39

UC: USS Monob YAG-61

UR: USS Interceptor YAGR-8

ML: USS Investigator YAGR Division 21

C:

MR: USS YFRT-520 Nuwes; Keyport, WA.

LL: USS Apopka YTB-778

LC:

LR: USS Forrest Royal

UL: USS Thuban

UC: USS Constellation

UR: USS James E. Keys

ML:

C:

MR:

LL: USS Sheapshead

LC: USS Eldorado

LR: USS Laffey

INDEX

Curts; FFG-38, 98
Cushing; DD-985, 66
Cusk; SS-348, 126
Cutlass; SS-478, 131

—D—

Dace; SSN-607, 145
Dahlgren; DDG-43, 72, 73
Dale; CG-19, 27
Dallas; SSN-700, 153
Damato; DD-871, 61
Daniels, Josephus; CG-27, 28
Daniels, Josephus; DLG-27, 85
Darter; SS-576, 134
Dash; MSO-428, 116
David, Albert; DE-1050, 77
David, Albert; FF-1050, 88
Davidson; DE-1045, 76
Davidson; FF-1045, 87
Davis; DD-937, 62
Davis, Rodney M.; FFG-60, 101
De Grasse, Comte; DD-974, 65
De Haven; DD-727, 54
De Soto County; LST-1171, 111
De Wert; FFG-45, 99
Decator; DDG-31, 71
Denver; LPD-9, 104
Detector; MSO-429, 116
Detroit; AOE-4, 13
Devastator; MCM-6, 115
Dewey; DDG-45, 73
Dewey; DLG-14, 84
Deyo; DD-989, 67
Diablo; AGSS-479, 131
Diamond Head; AE-19, 4
Diodon; SS-349, 127
Direct; MSO-430, 116
Dixie; AD-14, 1
Dixon; AS-37, 20
Dogfish; SS-350, 127
Dolphin; SS-555, 133
Dominant; MSO-431, 116
Dortch; DD-670, 52
Downes; DE-1070, 79
Downes; FF-1070, 90
Doyle; FFG-39, 98
Drum; SSN-677, 149
Dubuque; LPD-8, 104
Duluth; LPD-6, 104
Duncan; FFG-10, 95
Du Pont; DD-941, 63
Durham; LKA-114, 103
Duval County; LST-758, 110
Dyess; DD-880, 61

—E—

Eastman, George; YAG-39, 163
Edenton; ATS-1, 23
Edison, Thomas A.; SSBN-610, 135
Edson; DD-946, 63
Edwards, Richard S.; DD-950, 63
Eisenhower, Dwight D., CVN-69, 46
Elk River; IX-501, 101
Elliot; DD-967, 64
Ellison, Harold J.; DD-864, 60
El Dorado; Unknown, 164
El Paso; LKA-117, 103
El Rod; FFG-55, 100
Engage; MSO-433, 116
England; CG-22, 28
England; DLG-22, 84
Enhance; MSO-437, 116
Entemedor; SS-340, 126
Enterprise; CVAN-65, 43
Enterprise; CVN-65, 45, 50
Enwald; LSD-6, 107
Epping Forest; LSD-4, 107
Epping Forest; MCS-7, 115
Erben; DD-631, 51
Escape; ARS-6, 17
Essex; CVA-9, 41
Essex; CVS-9, 48
Estocin; FFG-15, 96
Evans, Frank E.; DD-754, 55
Eversole; DD-789, 56
Excel; MSO-439, 117
Exultant; MSO-441, 117

—F—

Fahrion; FFG-22, 96
Fairfax County; LST-1193, 114
Falgout; DER-324, 83
Fanning; DE-1076, 80
Fanning; FF-1076, 91
Farragut; DDG-37, 72
Farragut; DLG-6, 83
Fearless; MSO-442, 117
Fidelity; MSO-443, 117
Fife; DD-991, 67
Finback; SSN-670, 149
Firedrake; AE-14, 3
Fiske; DD-842, 59
Fitch, Aubrey; FFG-34, 98
Flagstaff; PGH-1, 119
Flasher; SS-249, 122
Flasher; SSN-613, 145
Flatley; FFG-21, 96
Fletcher; DD-992, 67
Flint; AE-32, 5
Florida; SSBN-728, 140
Florikan; ASR-9, 21
Flying Fish; SSN-673, 149
Ford; FFG-54, 100
Forrestal; CV-59, 36, 37
Forrest Royal; Unknown, 163
Forster; DER-334, 83
Fortify; MSO-446, 117
Fort Fisher; LSD-40, 109
Fort Mandon; LSD-21, 107
Fort Marion; LSD-22, 107
Fort McHenry; LSD-43, 109
Fort Snelling; LSD-30, 108
Foster; Paul F., DD-964, 64
Fox; CG-33, 29
Fox; DLG-33, 85
Fox; Myles C., DDR-829, 74
Franklin; CV-13, 35
Franklin, Benjamin; SSBN-640, 138
Frederick; LST-1184, 113
Fresno; LST-1182, 112
Frigate Bird; MSC-191, 115
Fulton; AS-11, 18
Furer, Julias A.; FF-6, 86
Furer, Julias A.; FFG-6, 94
Furse; DD-882, 62

—G—

Gainard; DD-706, 53
Gallant; MSO-489, 118
Gallery; FFG-26, 97
Gallup; PG-85, 118
Galveston; CLG-3, 34
Gansevoort; DD-608, 51
Garcia; FF-1040, 87
Garrett County; LST-786, 110
Gary; FFG-51, 99
Gates, Thomas S.; CG-51, 30
Gato; SSN-615, 146
Gemini; PHM-6, 120
Georgia; SSBN-729, 140
Germantown; LSD-42, 109
Gettysburg; CG-64, 32
Gilbert Islands; CVE-107, 44
Gilmore, Howard W.; AS-16, 19
Glennon; DD-840, 59
Glover; AGDE-1, 7
Glover; AGFF-1, 8
Glover; FF-1098, 93
Goldsborough; DDG-20, 70
Gompers, Samuel; AD-37, 2
Goodrich; DDR-831, 74
Graham County; AGP-1176, 8
Grand Canyon; AR-28, 16
Grand Rapids; PG-98, 119
Grant County; LST-1174, 112
Grant, Ulysses S.; SSBN-631, 138
Grapple; ARS-53, 18
Grasp; ARS-51, 18
Gray; DE-1054, 77
Gray; FF-1054, 88
Grayback; LPSS-574, 134
Grayling; SSN-646, 147
Great Sitkin; AE-17, 4
Green Bay; PG-101, 119
Green, Nathanial; SSBN-636, 138
Greene, Eugene A.; DDR-711, 74
Greenlet; ASR-10, 21
Greenling; SS-213, 121
Greenling; SSN-614, 146
Greenwood; DE-679, 75
Gregory; DD-802, 57
Gridley; CG-21, 28
Groton; SSN-694, 152
Groves, Stephen W.; FFG-29, 97
Growler; SSG-577, 134
Guadalcanal; LPH-7, 106
Guadalupe; A0-32, 10
Guam; LPH-9, 106
Guardfish; SSN-612, 145
Gudgeon; SS-507, 132
Gudgeon; SS-567, 133
Guitarro; SSN-665, 148
Gurke; DD-783, 56
Gurnard; SSN-662, 148

—H—

Haddo; SSN-604, 145
Haddock; SSN-621, 146
Hale; DD-642, 52
Hale, Nathan; SSBN-623, 137
Haleakala; AE-25, 4
Halfbeak; SS-352, 127
Halibut; SSGN-587, 142
Halibut; SSN-587, 143
Hall, Gunston; LSD-5, 107
Hall, Gunston; LSD-44, 109
Hall, John L.; FFG-32, 97
Halsey; CG-23, 28
Halyburton; FFG-40, 98
Hamilton, Alexander; SSBN-617, 136
Hammerhead; SSN-663, 148
Hammond, Francis; DE-1067, 78
Hammond, Francis; FF-1067, 90
Hamner; DD-718, 54
Hamul; AD-20, 2
Hancock; CVA-19, 41, 42
Hancock, John; DD-981, 66
Hank; DD-702, 53
Hansford; APA-106, 14
Harder; SS-568, 133
Hardhead; SS-365, 127
Harkness; T-AGS-32, 161
Harlan County; LST-1196, 114
Hart, Thomas C.; DE-1092, 82
Hart, Thomas C.; FF-1092, 93
Hassayampa; AO-145, 11
Haverfield; DER-393, 83
Hawes; FFG-53, 99
Hawkbill; SSN-666, 148
Hawkins; DD-873, 61
Hayes; T-AGOR-16, 161
Hayler; DD-997, 67
Hector; AR-7, 15
Helena; SSN-725, 158
Henderson; DD-785, 56
Henley; DD-762, 55
Henry, Patrick; SSBN-599, 135
Hepburn; FF-1055, 88
Hercules; PHM-2, 120
Hermitage; LSD-34, 108
Hewes, Joseph; DE-1078, 80
Hewes, Joseph; FF-1078, 91
Hewitt; DD-966, 64
Higbee; DD-806, 57
High Point; PCH-1, 118
Hill, Harry W.; DD-986, 66
Hitchiti; ATF-103, 22
Hoel; DDG-13, 69
Hoist; ARS-40, 18
Holder; DD-819, 57
Holland; AS-32, 19, 20
Hollidaysburg; PCS-1385, 118
Hollister; DD-788, 56
Holt, Harold E.; DE-1074, 79
Holt, Harold E.; FF-1074, 90
Honolulu; SSN-718, 157
Hooper; DE-1026, 76
Horne; CG-30, 28
Horne; DLG-30, 85
Hornet; CVS-12, 48
Houston; SSN-713, 156
Houston; Sam; SSBN-609, 135
Hubbard, Harry E.; DD-748, 55
Hull; DD-945, 63
Hunley; AS-31, 19
Hunterdon County; LST-838, 110
Huntington, R.K.; DD-781, 56
Huse; DE-145, 75

—I—

Icefish; SS-367, 127
Illusive; MSO-448, 117
Implicit; MSO-455, 117
Inchon; LPH-12, 106
Independence; CV-62, 38
Indianapolis; SSN-697, 153
Inflict; MSO-456, 117
Ingersoll; DD-652, 52
Ingersoll; DD-990, 67
Ingraham; DD-694, 53
Ingraham; FFG-61, 101
Ingram, Jonas; DD-938, 62
Intercepter; YAGR-8, 163
Intrepid; CVS-11, 48
Investigator; YAGR, 163
Iowa; BB-61, 24

Paul; FF-1080, 91
Peary, Robert E.; DE-1073, 79
Peary, Robert E.; FF-1073, 90
Pegasus; PHM-1, 119
Peleliu; LHA-5, 102
Pennsylvania; SSBN-735, 142
Pensacola; LSD-38, 109
Peoria; LST-1183, 113
Perch; APSS-313, 124
Permit; SS-178, 120
Permit; SSN-594, 144
Perry, John R.; DE-1034, 76
Perry, N.K.; DD-883, 62
Perry, Oliver Hazard; FFG-7, 94
Peterson; DD-969, 64
Petrel; ASR-14, 21
Pharris; FF-1094, 93
Phelps; DD-360, 50
Philadelphia; SSN-690, 151
Philip, George; FFG-12, 95
Philippine Sea; CG-58, 31
Philippine Sea; CV-47, 36
Phoenix; SSN-702, 153
Pickaway; APA-222, 15
Picking; DD-685, 52
Piedmont; AD-17, 1
Pigeon; ASR-21, 22
Pintado; SSN-672, 149
Pirana; SS-389, 128
Pittsburgh; SSN-720, 157
Plainview; AGEH-1, 8
Platte; AO-186, 12
Pledge; MSO-492, 118
Pluck; MSO-464, 117
Plunger; SSN-595, 144
Plymouth Rock; LSD-29, 108
Pogy; SSN-647, 147
Point Defiance; LSD-31, 108
Point Loma; AGDS-2, 7
Polk, James K.; SSBN-645, 139
Pollack; SS-180, 120
Pollack; SSN-603, 145
Pomfret; SS-391, 128
Ponce; LPD-15, 105
Ponchatoula; AO-148, 12
Porpoise; SS-172, 120
Portland; LSD-37, 109
Portsmouth; SSN-707, 155
Power; DD-839, 59
Prairie; AD-15, 1
Pratt, William V.; DDG-44, 73
Pratt, William V.; DLG-13, 84
Preble; DDG-46, 73
Preble; DLG-15, 84
Preserver; ARS-8, 17
Prichett; DD-561, 51
Princeton; CG-59, 31
Princeton; CVS-37, 49
Proteus; AS-19, 19
Providence; SSN-719, 157
Puffer; SSN-652, 147
Puget Sound; AD-38, 2
Pulaski, Cassimir; SSBN-633, 138
Puller, Lewis B.; FFG-23, 96
Pyro; AE-24, 4

—Q—

Queenfish; SSN-651, 147

—R—

Racine; LST-1191, 114
Radford, Arthur W.; DD-968, 64
Raleigh; CL-7, 33
Raleigh; LPD-1, 104
Ramsey; DEG-2, 82
Ramsey; FFG-2, 94
Randolph; CVA-15, 41
Randolph; CVS-15, 49
Ranger; CV-61, 37, 38
Ranger; CVA-61, 43
Rankin; AKA-103, 9
Rasher; SS-269, 122
Rathburne; FF-1057, 88
Ray; SSN-653, 147
Ray; SSR-271, 123
Ray, David R.; DD-971, 65
Rayburn, Sam; SSBN-635, 138
Reasoner; DE-1063, 78
Reasoner; FF-1063, 89
Reclaimer; ARS-42, 18
Redfish; SS-395, 128
Redstone; T-AGM-20, 161
Reeves; CG-24, 28
Reeves; DLG-24, 84
Reid; FFG-30, 97
Remey; DD-688, 52
Remora; SS-487, 132
Rentz; FFG-46, 99
Requin; SS-481, 131
Requin; SSR-481, 132
Resolute; AFDM-10, 6
Revere, Paul; LPA-248, 103
Rich; DD-820, 58
Richard, Bon Homme; CVA-31, 42
Ricketts, Claude V.; DDG-5, 68
Rickover, Hyman G.; SSN-709, 155
Rigel; AF-58, 6
Rigel; T-AF-58, 160
Rivers, L. Mendel; SSN-686, 151
Roan, Charles H.; DD-853, 60
Roanoke; AOR-7, 14
Roanoke; CL-145, 34
Roark; DE-1053, 77
Roark; FF-1053, 88
Roberts, Samuel B.; DD-823, 58
Roberts, Samuel B.; FFG-58, 101
Robison; DDG-12, 69
Rodgers, John; DD-983, 66
Rogers; DD-876, 61
Rogers, Will; SSBN-659, 140
Ronquil; SS-396, 128
Roosevelt, Franklin D.; CVA-42, 43
Roosevelt, Theodore; CVN-71, 47, 48
Roosevelt, Theodore; SSBN-600, 135
Ross; DD-563, 51
Rowan; DD-405, 50
Runner; SS-476, 131
Rupertus; DD-851, 60
Rush, William R.; DD-714, 53
Rush, W.R.; DDR-714, 74
Russell, Richard B.; SSN-687, 151

—S—

Sabalo; SS-302, 123
Sacramento; AOE-1, 12
Safeguard; ARS-25, 17
Safeguard; ARS-50, 18
Saginaw; LST- 1188, 113
Sailfish; SS-572, 133
Saint Louis; LKA-116, 103
Saint Paul; CA-73, 26
Saipan; LHA-2, 102
Salmon; SS-573, 133
Salt Lake City; SSN-716, 157
Salute; MSO-470, 118
Salvor; ARS-52, 18
Sample; DE-1048, 77
Sample; FF-1048, 87
Sampson; DDG-10, 68, 69
San Bernardino; LST-1189, 113
San Diego; AFS-6, 7
San Francisco; SSN-711, 155
San Jacinto; CG-56, 31
San Jacinto; CVL-30, 45
San Jose; AFS-7, 7
San Juan; SSN-751, 158
San Marcos; LSD-25, 107
San Onofre; ARD-30, 16
Sanctuary; AH-17, 9
Sand Lance; SS-381, 127
Sand Lance; SSN-660, 147
Santa Barbara; AE-28, 5
Saratoga; CV-3, 34
Saratoga; CV-60, 37
Saratoga; CVA-60, 43
Sarda; SS-488, 132
Sargo; SSN-583, 143
Savage; DER-386, 83
Savannah; AOR-4, 13
Sawfish; SS-276, 123
Scamp; SSN-588, 143
Schenectedy; LST-1185, 113
Schofield; DEG-3, 82
Schofield; FFG-3, 94
Scorpion, SSN-589, 143
Scott; DDG-995, 73
Sculpin; SSN-590, 144
Sea Cat; SS-399, 129
Sea Cliff; DSV-4, 86
Sea Devil; SS-400, 129
Sea Devil; SSN-664, 148
Sea Dog; SS-401, 129
Seadragon; SSN-584, 143
Sea Fox; SS-402, 129
Sea Horse; SSN-669, 148
Sea Owl; SS-405, 129
Sea Robin; SS-407, 129
Sea Wolf; SSN-575, 142
Seattle; AOE-3, 13
Segundo; SS-398, 129
Sellers; DDG-11, 69
Semmes; DDG-18, 69
Sentry; MCM-3, 115
Shadwell; LSD-15, 107
Shangri-La; CVA-38, 42
Shangri-La; CVS-38, 49
Shark; SSN-591, 144
Shasta; AE-6, 3
Shasta; AE-33, 5
Sheepshead; Unknown, 164
Shelton; DD-790, 57
Shenandoah; AD-26, 2
Shenandoah; AD-44, 3
Sherman, Forrest; DD-931, 62
Shields, Marvin; DE-1066, 78
Shields, Marvin; FF-1066, 89, 90
Shippingport; ARDM-4, 17
Shreveport; LPD-12, 105
Sides; FFG-14, 96
Sierra; AD-18, 1
Silversides; SS-236, 121
Silversides; SSN-679, 150
Simpson; FFG-56, 100
Sims, W.S.; DE-1059, 78
Sims, W.S.; FF-1059, 89
Sirius; T-AFS-8, 160
Skate, SSN-578, 142
Skipjack; SSN-585, 143
Skylark; ASR-20, 21
Small, Ernest G.; DDR-838, 74
Snapper, SS-185, 121
Snohomish County; LST-1126, 110
Snook; SSN-592, 144
Somers; DDG-34, 71
South Carolina; CGN-37, 33
South Dakota; BB-57, 24
Southerland; DD-745, 54
Spadefish, SSN-668, 148
Spartanburg County; LST-1192, 114
Spear, L.Y.; AS-36, 20
Spearfish; SS-190,121
Sperry; AS-12, 18
Sperry, Chas S.; DD-697, 53
Sphinx; ARL-24, 17
Spiegel Grove; LSD-32, 108
Spinax; SSR-489, 132
Spot; SS-413, 130
Sprague, Clifton; FFG-16, 96
Spraunce; DD-963, 64
Springfield; CLG-7, 34
Standley, William H.; CG-32, 29
Standley, William H.; DLG-32, 85
Stark; FFG-31, 97
Stark County; LST-1134, 111
Steadfast; AFDM-14, 6
Stein; DE-1065, 78
Stein; FF-1065, 89
Steinaker; DD-863, 60
Sterett; CG-31, 29
Sterlet; SS-392, 128
Steuben, Von; SSBN-632, 138
Stickleback; SS-415, 130
Stimson, Henry L.; SSBN-655, 139
Stingray; SS-186, 121
Stoddard; DD-566, 51
Stoddert, Benjamin; DDG-22, 70
Stone County; LST-1141, 111
Strauss, Joseph; DDG-16, 69
Stribling; DD-867, 61
Strong; DD-758, 55
Stump; DD-978, 65
Sturgeon; SS-187, 121
Sturgeon; SSN-637, 146
Suffolk County; LST-1173, 111
Sumner, Allen M.; DD-692, 52
Sumter; LST-1181, 112
Sunbird; ASR-15, 21
Sunfish; SSN-649, 147
Suribachi; AE-21, 3
Sustain; AFDM-7, 6
Swenson, Lyman K.; DD-729, 54
Swordfish; SSN-579, 143
Sylvania; AFS-2, 6

—T—

Tacoma; PG-92, 119
Takelma; ATF-113, 23
Talbot; DEG-4, 82
Talbot; FFG-4, 94
Tallahatchie County; AVB-2, 24
Tallahatchie County; LST-1154, 111
Tang; SS-563, 133
Tarawa; LHA-1, 102
Tattnall; DDG-19, 70
Taurus; PHM-3, 120

582
BONEFISH

Printed in the USA
CPSIA information can be obtained
at www.ICGtesting.com
JSHW060054150824
68134JS00032B/2727

9 781630 269227